YOUR KNOWLEDGE HAS VALUE

- We will publish your bachelor's and master's thesis, essays and papers

- Your own eBook and book - sold worldwide in all relevant shops

- Earn money with each sale

Upload your text at www.GRIN.com
and publish for free

Bibliographic information published by the German National Library:

The German National Library lists this publication in the National Bibliography; detailed bibliographic data are available on the Internet at http://dnb.dnb.de .

This book is copyright material and must not be copied, reproduced, transferred, distributed, leased, licensed or publicly performed or used in any way except as specifically permitted in writing by the publishers, as allowed under the terms and conditions under which it was purchased or as strictly permitted by applicable copyright law. Any unauthorized distribution or use of this text may be a direct infringement of the author s and publisher s rights and those responsible may be liable in law accordingly.

Imprint:

Copyright © 2018 GRIN Verlag
Print and binding: Books on Demand GmbH, Norderstedt Germany
ISBN: 9783668894006

This book at GRIN:

https://www.grin.com/document/453166

Sven Meyer

The future usage of wood. Timber as a sustainable material in construction

GRIN Verlag

GRIN - Your knowledge has value

Since its foundation in 1998, GRIN has specialized in publishing academic texts by students, college teachers and other academics as e-book and printed book. The website www.grin.com is an ideal platform for presenting term papers, final papers, scientific essays, dissertations and specialist books.

Visit us on the internet:

http://www.grin.com/

http://www.facebook.com/grincom

http://www.twitter.com/grin_com

Timber as a sustainable material in construction: An investigation of the future usage of wood and the indicators that can influence a change

Sven Meyer
Worcester Polytechnic Institute

Table of contents

Abstract 3
1. Introduction 4
2. Resources and Market 4
3. Transportation of Timber 5
4. Comparison to other building materials 7
5. Estimation of Wood Usage 9
6. Conclusion 10
References 12

Abstract

Wood, the only truly sustainable construction material, should play a key role when it comes to future applications. Current developments of different construction methods enable new applications of wooden elements in buildings. The increasing popularity of sustainability requires an investigation of the indicators that can predict a change in the future, which is the main goal of this paper, specialized to the U.S. construction industry. To this end, the fundamental influences were examined. The U.S. housing market shows an increasing demand, which leads to the consumption of more construction material. The managed forest areas in the U.S. must be increased to meet the future needs and preserve wood as a sustainable material. This applies not only to the U.S. It is an important consideration for countries the U.S. depends on, which is primarily Canada. Further research and developments have to be made on the prefabrication and material issues, like fire resistance and weather protection, in order to make timber a sustainable, advantageous material in the construction field. An estimation of the required timber resources for a building was examined to provide further investigations, not only to estimate the need of new building materials in the future, but also the required forest resources whether it is in terms of import or the usage of the U.S.' own resources.

Keywords: Construction, Environment, Estimation, Forest, Resources, Sustainability, Timber, U.S., Wood

Acknowledgement

The author would like to thank Professor Leonard D. Albano of the Civil & Environmental Engineering Department from the Worcester Polytechnic Institute.

1. Introduction

Thinking about the environment, the meaning of sustainability could be defined as a natural resource balance. The main principle can be claimed as balancing the needs of our current generation without any negative impact on our future generations. This means sustainable projects are those which do not cause irreparable damage to the environment. (Li, et al, 2017) Sustainability is a common topic in this time and age, a lot of trends about sustainable developments emerging within the last years in different areas. Also, the province of construction is more focusing on sustainability with already established certificates, for instance Leadership in Energy and Environmental Design (LEED) or German Sustainable Building Council (DGNB). Thinking about materials in construction, different areas of application could be considered. The main aspect, which this paper will focus on, is the construction material timber. Not only does sustainability not receive as much attention in the construction industry as in other fields, but also the construction methods based on wood. Through the engineering development of wooden products for construction, more efficient and economically more beneficial solutions could appear on the market. (Švajlenka, J. & Kozlovská, M. 2017)

This paper will at first focus on the wood resources in the U.S. and the recent developments in the U.S. housing market. Through that, the importance of sustainable timber construction will be shown and whether sufficient resources are available to keep up with future needs. The import and export of the wood resources has to be considered here to understand why both the import and export of timber are essential for the U.S. Furthermore, the importance of timber as a sustainable material will be shown by comparison to other building materials. Weighing the pros and cons of timber as a sustainable material compared to others, the implementation in the context of taller buildings and fire resistance will be discussed. Economic growth lets high buildings in big cities occur; however, timber as a structural material can barely be discovered on such tower buildings. In order to make wood more sustainable, prefabrication is another topic that stands out and hasn't been developed that far to date. At the end there will be a factor provided that helps further estimations when it comes to the usage of wood in buildings.

2. Resources and Market

The U.S. is among the top countries regarding wood resources; it holds 8% of the world's forest area, which is number 4 on the global list, right below Canada. The protected forest areas in the U.S. make up 11% of the protected forests in the world. Additionally, the U.S. gained an average 0.1% annually of forest area between 2010-2015. They reached number 2 among the countries regarding wood removal in 2011, whereof 87.5% of the harvested wood was used for different kinds of production. Through the increased usage of timber in construction projects, several points must be considered. One is deforestation. In order to heed this potential problem, the wood needs to be harvested from specific areas within the forest environment so that the forests will not be incrementally damaged. (Ramage, et al. 2017) Such areas are multiple-use managed forest areas, under which the wood used for construction purposes counts. (UNECE, 2015) The WWF suggests that 25% more of the managed wood areas should be developed by 2050, including in the U.S. in order to avoid a negative impact on the environment through the harvesting of forest areas. Therefore, better managed and larger forest areas are needed; this also includes the involvement of the government with the help of law enforcement to better regulate the trade and the supply chains. (WWF, 2015)

The construction market has an overall growth. One reason for that are the investments of the U.S. on single-family housings and multi-family housings in 2016. The multi-family housing has the biggest growth in investments by 24%. Single-family housings increased by approximately half of that. One reason is that it is simply less expensive per unit to construct a multi-family house instead of several single-family houses. The increasing numbers in the real estate market show the importance of sustainability and the choice of the building material. (Buehlmann, et al. 2017)

Unlike other materials such as gravel or sand, timber is categorized as a renewable material. The implementation of a higher usage of timber shows only benefits in terms of being renewable and sustainable if the growth and the deforestation are balanced. One important issue here to consider is the natural growth rate of wood, which is why the plantation of new trees needs to be sustainable with a glance into the future behavior in terms of our constructional movements and the involvement of timber in it. Another issue to take into account is that mostly softwood is used for construction, especially spruce and pine. This means that the resources of these special kinds of wood needs to be investigated in order to ascertain the opportunity for further use of timber in the construction field. (Pohl, 2018)

3. Transportation of Timber

Timber is special regarding transportation. Whereas other construction materials such as cement, aggregates, and metals are mechanically extracted from the ground. Timber needs topsoil and seedlings to grow before it can be used for special purposes. The growth time trees need prior to harvesting varies between 35 to 70 years, depending on the species and the location of the forest area. For the reason that trees are growing, and other construction materials are rather existent and endemic, timber is the only real sustainable material in the field of construction. (Ramage, et al. 2017)

Timber as a wood-based product is derived from round wood. Most of the world's round wood plantations are based in Canada and China, which shows that the transportation of timber has to be involved to supply other countries. (Lovarelli, et al.2018) The import index of wooden products increased by approximately 40% from 2005 to 2018. Something that has to be considered here is that the index increased by 10% in a period of 10 years, i.e. the index raised 30% within the last 3 years alone. Therefore, the dependence on other countries e.g. Canada for the import of wooden products, like timber will play a key role in the future. (BLS, 2018)

The problem for many countries is that they require an import of wood from resources because of their lack of managed forest areas. For example, wood import to Taiwan was investigated. Among several suppliers were the U.S. and Canada. To understand the issues of the import, different stages have to be considered. At first, it comes to the harvesting of the wood, and this step requires mechanical tools like a saw. The next step is the transportation from the forest region to a sawmill. In order to keep the transportation costs low, the sawmill shouldn't be located too far from the harvesting location. Subsequently the wood gets manufactured in the sawmill. Besides processes like chipping, processes like drying and gluing are included here. Next the manufactured wood has to be shipped to the port. By doing this the wood can be transported to Taiwan. Herein the average distances within the U.S. differ depending on the region. The lowest distance is the sawmill in Oregon or Washington to the port in Tacoma with 110km. The biggest distance are sawmills in the Southeast to the port in Houston with 919km, whereas Canada has an average distance from their sawmills to their port of 555km.

The import of wood from Canada and other countries demands lower amount of energy than it does from the U.S. Here the U.S. needs more than double the energy compared to Canada because the energy used for the manufacturing is much lower in Canada. Besides the required energy, also more embodied CO_2 emissions are coming along with the process in the U.S., whereof Canada has half of the numbers. This is the reason why the U.S. gets a lot of wood delivered from Canada. The energy costs of the supply chains from the manufacturing to the import in another country are significantly lower. (Li, et al, 2017)

Within the last decade, hardwood export increased continuously and doubled the million Dollar worth in less than 10 years, although there are ups and downs between single years. Approximately half of all hardwood exports are shipped to China to date. A decade ago, the situation was totally different, and they just received even less wood exports as Canada which makes an-eighth by now. This number hasn't changed that much within the last years. Countries like Mexico and Vietnam also increased their wood import from the US. (Buehlmann, et al. 2017)

Figure 1 shows clearly the huge discrepancy between the export and import price for the U.S. The export of their own forest resources is more profitable and economically more rational than the usage of it for own construction purposes. One reason for that is the low-cost import from Canada, which allows a successful trade between borders. What shows that the import of timber from Canada, because of the reasons mentioned above, is more sustainable than using their own resources. Moreover, the export of their sources brings economic growth.

Fig.1: U.S. Roundwood Unit Price. (UNECE 2018)

4. Comparison to other building materials

Properly covered and protected from the elements, timber can have the same lifespan as other structural materials. A study by the Athena Institute showed that most buildings do not have problems with the structural material (The Athena Institute, 2004). From 227 investigated demolished buildings, there were only 8 of them demolished because of the structural material; moreover 27 wooden buildings which were older than 100 years haven't had a problem with the structural material at all. Based on the application area the characteristics of timber can be useful or likewise not useful. The density is low compared to other common materials like concrete and steel. Another attribute of timber is the lower stiffness. The strength of hardwood, parallel to the grain, is even a bit stronger compared to that of concrete, whereas softwood is a bit weaker. In the case strength of high strength concrete, wood isn't able to keep up regarding the compression here. These characteristics qualify the usage of wood for tall structures, where the main part of the load, which is its own weight, can be carried by the structure. That's why wood is often used for roofs, but also for bridges or as part of systems in tall buildings. Wood as a structural material can't be used as the only structural material in tall buildings though. The load would be too large just for wood to carry. There have been several systems developed in the last decades, which allow a combination even if there are always consequences because the usage of a combination of two different materials as part of the structural system. The connections for the combination of different materials is more widely developed across other materials. Although there are working systems with wood, still a lot of research must be conducted on that field.

The selection of the different structural systems for buildings depends on the height and the kind of implantation of the wood. Up to six stories, the use of a light timber frame is most beneficial in terms of the amount of timber used. Buildings from six to ten stories should use cross-laminated timber (CLT) in combination with a light timber frame to ease the amount of timber used. With more than ten stories the integration of wood in buildings can only use cross-laminated timber units. (Ramage, et al. 2017)

The material wood is one of the less energy-intensive materials used for construction; therefore, the enhanced usage of wood would help to reduce the greenhouse gas emission through the lowering of high energy-intensive materials like iron and concrete. (UNECE, 2015) Timber-based systems require around 15% less energy in terms of non-heating purposes compared to steel or concrete-based systems. One item that has to be mentioned here is that the greenhouse gas benefits are much more significant than the energy benefits when it comes to the substitution of the main construction materials through the wood. Timber-based houses create between 20-50% fewer emissions. This is a significant number which has to be pointed out when thinking about future investments in building construction and the impact of construction on our environment. (Upton, et al. 2007)

Another point is moisture, and the weather contributes to moisture. During construction; timber must be covered against occurring weather issues. Otherwise there will be delays, problems during the work, and additional costs. Moisture damages through missing protection is a common problem on sites, and such problems could be avoided through an in-transit-protection of the timber elements and additional protection on the site itself. (Ruuska, Antti & Häkkinen, Tarja, 2016) Moisture causes a change of mechanical properties and dimensions and the impact varies with the level of moisture. Wood, as any other material, needs maintenance and has a limited durability. Another point to consider here is that a high level of moisture allows insects and fungi to demolish the material. That is why the design should avoid wetting and direct sunlight as much as possible. One way to do this is to raise columns above overhanging roofs and the ground level; such little considerations can ensure to last the structural material for centuries. (Ramage, et al. 2017)

There are two main differences to consider using timber in construction. On one side there is the prefabricated building method, which owns more than 80% of the market and clearly became the standard method. On the other side is the conventional building method. (Pohl, 2018) Industrial manufacturing is currently a problem within timber production that other fields in construction doesn't have. In spite of the fact that it improves the quality, there need to be improvements done in terms of the manufacturing costs and schedule benefits. The faster installation of prefabricated elements shows its benefits especially when it comes to large buildings, including timber. (Ruuska, Antti & Häkkinen, Tarja, 2016)

New wooden products, which emerged from research and development, have a strong influence on the construction market. Wood products, like the above mentioned CLT, also called "mass timber" or thermally modified wood (TMW) support the construction process and the work of architects and engineers through their attributes, such as durability, stability, and fire resistance. Generally, CLT and TMW products are prefabricated which makes them easy to install, enabling shorter construction periods. TMW can last in high temperatures up to 260°C for a short period of time. The good strength-to-weight ratio of CLT allows the implementation of buildings for static purposes. (Buehlmann, et al. 2017) The United States Department of Agriculture (USDA) sponsored a tall wood building competition, where they awarded among another one, a twelve-floor building in Portland that mixed an office and residential building. Here the building used CLT floors and glued laminated timber for the structural frame. Such projects show the successful implementation of timber in modern construction, especially when supported by the U.S. government. Through such support, faster improvement can be reached in a shorter period.

The fire resistance of timber must be sufficient to enable the people inside the building to safety evacuate. A usually required fire resistance for timber would be 60-120 minutes, which mostly depends on the height of the building and the connectors to the timber. (Barber, 2017) Timber suffers a reduction in its stiffness and strength at a lower temperature than steel or concrete. For example, timber loses 50% of the stiffness and strength at 100°C compared to its stiffness and strength at 20°C. The fire resistance of timber needs further research for taller buildings that include wood. (Ramage, et al. 2017)

5. Estimation of Wood Usage

To measure the usage of wood in construction, a coefficient will be used that shows the approximate amount of timber per 1000m^3 of enclosed volume. This amount is around 17m^3, although this value is getting much higher in massive timber construction buildings, where it is 90m^3 per 1000m^3 enclosed volume. (Pohl, 2018) The median size is around 470m^2 for a typical commercial building i.e. approximately half of the buildings are smaller than this amount of m^2 and the half of buildings are bigger than this amount. (CBECS, 2015)

Using this estimate of timber usages, it is possible to estimate the average amount of wood used in commercial buildings with the help of the coefficients. In order to consider enclosed volume, we assume an average height of 2.75m. (Bernstein, 2006) Therefore, the number used for the calculation would be 1300m^3. The median house size in the U.S. is around 230m^2, what can vary year to year, whereby the enclosed volume in m^3 would be about 630m^3. (Perry, 2016) Although the average house size varies every year, this number can be decreased by the increasing number of constructed multi-family houses, which usually provide a smaller living size of commercial buildings to be as profitable and beneficial as possible. Therefore, this number needs adjustment among the time, in order to make sure the estimation is still valid.

Table 1: Timber Construction Buildings
Timber Construction Buildings

Building Type	Size	m^3 Wood per 1000 m^3	Used Wood
Housing Building	630m^3	17	**10,71m^3**
Commercial Building	1300m^3	17	**22,10m^3**

Table 2: Massive Timber Construction Buildings
Massive Timber Construction Buildings

Building Type	Size	m^3 Wood per 1000 m^3	Used Wood
Housing Building	630m^3	90	**56,70m^3**
Commercial Building	1300m^3	90	**117,00m^3**

These numbers can be used to calculate an estimation of the required material for wooden buildings with the amount of planned buildings in a country, state or municipality. The import prices can be used with the calculations shown above in order to estimate the cost of the needed timber. Furthermore, future estimations can be done with the consideration of managed forest resources, which helps to keep the importance of managed forest areas in mind and how much wood can be involved in future construction investments. Such numbers can be established for the different timber combination systems to obtain an even more accurate number. Through that, the cost estimation of timber included buildings can be calculated easier.

The following calculation shows an example application of the coefficients. In the year 2015, 99,800,000m^3 round wood was imported to the U.S. (Statista, 2018) In the same year, 628,000 single-houses were built. (U.S. Department of Commerce, 2015) In typical timber construction buildings, the required wood for these houses would have been about 6,725,800m^3, which is 6.7% of the imported round wood in 2015. The amount of required wood for all built single-houses in the U.S. would have cost around half a billion dollars based on the import index of round wood. An additional consideration is the kind of imported wood. Manufactured wood products are more cost-ineffective: wood-based panels are 6 times more cost-intensive compared to round wood. Plywood would be 8 times higher. (UNECE, 2018) The distinction of different kinds of imported wood, shows that already manufactured wood is more cost-intensive.

6. Conclusion

Through the investigation of the forest resources in the U.S., the importance of managed forest areas became clear. Although the U.S. is one of the top countries in terms of forest resources, a lot of those resources are nationally protected areas. Other areas might be inadequate for the manufacturing of timber. Multi-use forest areas, which are dedicated to the production of wooden products, must be increased and managed in the future in order to prevent a lack of resources, which would make wood no longer a sustainable construction material according to the described definition in this paper. The growing U.S. housing market shows the future need of new accommodations. Therefore, the sustainability should be crucial here. The increased, controlled usage of wood as a building material can develop a sustainable future in the construction province. Like every material, wood has to be transported from the harvested area. Wood is very special in terms of transportation and national as well as international, transportation is an essential aspect when it comes to sustainability, which needs closer investigations. It became apparent that the import of timber from Canada is more economical than the usage of own wood resources. One main cause is that the export of timber e.g. to Taiwan is more likely to be lucrative. Additionally, the import of wood is significantly cheaper, especially from Canada, this is because of the lower manufacturing costs in Canada. The dependence on Canada for lower cost timber is therefore substantial. Without the resource support of Canada, the usage of wood in construction would decrease because of the increasing unit price for timber, that is what makes alternative common materials more profitable.

The comparison to other construction materials showed the main advantages and disadvantages of timber. An important future market is the implementation of timber in tall buildings. Although wood can't be used as the only structural material in higher buildings, it can be used as a part of the structural material with the combination of common materials like steel and concrete. The development of new prefabricated timber elements allow a faster installation on the site with cost and time benefits. The prefabricating process though requires further development. Currently prefabricated timber accounts for 80% of the market but the costs of the production itself are more intensive, and wood components take a longer lead time than other prefabricated products. Other problems that need further development and research in the following years are fire resistance and moisture. These are challenges to widespread use of timber as the sustainable material with all its benefits like the significant lower greenhouse gas emissions during the fabrication. New types like the TMW allows an improvement of the fire resistance but still can't reach the fire resistance of other materials. The problem of moisture can be avoided through the appropriate usage in the building and the consideration of this problem before it gets stored on the site, as well as during transportation. This can be supported through regular processes in the supply chain with a clear structure.

The added estimation of wood usage allows future calculations for buildings, which helps to evaluate timber as a sustainable material in terms of the required resources. In summary it is essential to say that the future usage of timber as a construction material, that follows up with the idea of sustainability, can be increased in the U.S. and other parts of this world through new methods of implementing timber in buildings. This needs further research in different fields. The resources of managed wood areas need optimization to ensure the steady supply of this material in the future. Further projects of wood usage can inform the involvement of the government and may accelerate the process.

References

Barber, David (2017) Determination of fire resistance ratings for glulam connectors within US high rise timber buildings. *Fire Safety Journal*. [Online] 91579–585

Bernstein, Fred (2006) Developers and Architects Face a Tall Order From Buyers. The New York Times. https://www.nytimes.com/2006/01/22/realestate/developers-and-architects-face-a-tall-order-from-buyers.html

BLS (2015-18) Bureau of Labor Statistics *U.S. Import and Export Price Indexes;2015-201*. [online]. Available from: https://www.bls.gov/news.release/ximpim.toc.html

Buehlmann, Urs et al. (2017) Recent Developments in US Hardwood Lumber Markets and Linkages to Housing Construction. *Springer International Publishing*. [Online] 213-222

CBECS (2015) Commercial Buildings Energy Consumption Survey: A Look at the U.S. Commercial Building Stock: Results from EIA's 2012 Commercial Buildings Energy Consumption Survey. US Energy Information Administration. Available from: https://www.eia.gov/consumption/commercial/reports/2012/buildstock/

Li, S. et al. (2017) Identifying sustainable wood sources for the construction industry: A case study. *Sustainability (Switzerland)*. [Online] 10 (1),

Lovarelli, Daniela et al. (2018) Delving the environmental impact of roundwood production from poplar plantations. *Science of the Total Environment*. [Online] 645646–654.

Perry, Mark (2016). New US homes today are 1,000 square feet larger than in 1973 and living space per person has nearly doubled. American Enterprise Institute. Available from: http://www.aei.org/publication/new-us-homes-today-are-1000-square-feet-larger-than-in-1973-and-living-space-per-person-has-nearly-doubled/

Pohl, Sebastian. (2018). When green marketing meets reality: selected facts about sustainable house building. *Mauerwerk*. [Online] 22 (4), 215–224.

Ramage, Michael H. et al. (2017) The wood from the trees: The use of timber in construction. *Renewable and Sustainable Energy Reviews*. [Online] 68333–359.

Ruuska, Antti & Häkkinen, Tarja (2016) Efficiency in the Delivery of Multi-Story Timber Buildings. *Energy Procedia*. [Online] 96190–201

Statista, (2018). Total United States industrial roundwood imports and exports from 2006 to 2015. *The Statistics Portal*. [Online]. Available from: https://www.statista.com/statistics/252705/total-us-industrial-roundwood-imports-and-exports-since-2001/

Švajlenka, J. & Kozlovská, M. (2017) Perception of user criteria in the context of sustainability of modern methods of construction based on wood. *Sustainability (Switzerland)*. [Online] 10 (2),

The Athena Institute. (2004) *Minnesota Demolition Survey. Technical report, The Athena Institute*. [Online]

Upton, Brad et al. (2007) The greenhouse gas and energy impacts of using wood instead of alternatives in residential construction in the United States. *Biomass and Bioenergy*. [Online] 32 (1), 1–10.

UNECE (2015) Global Forest Resources Assessment 2015: How are the world's forests changing? *Forest Ecology and Management*. [Online]. Available from: http://www.unece.org/forests/areas-of-work/forest-resources.html

UNECE (2018) Export and Import Unit Prices (1964-2017). *United Nations Economic Commission for Europe* [Online]. Available from: http://www.unece.org/forests/output/prices.html

U.S. Department of Commerce. (2015). 2015 Characteristics of new Housing. *United States Census Bureau*. Available from: https://www.census.gov/construction/chars/pdf/c25ann2015.pdf

WWF (2015) WWF Living Forest Report: Chapter 5 Saving Forests at Risk. *World Wildlife Fund*. [Online]. Available from: https://www.wwf.de/fileadmin/fm-wwf/Publikationen-PDF/WWF-Living-Forests-Report-Chapter-5.pdf

YOUR KNOWLEDGE HAS VALUE

- We will publish your bachelor's and master's thesis, essays and papers

- Your own eBook and book - sold worldwide in all relevant shops

- Earn money with each sale

Upload your text at www.GRIN.com
and publish for free